Introduct
ion

Monkeys and gorillas are the wardrobe relative in the animals of the world collectively to people. A considerable lot of the species share up to 98 percent of similar DNA as individuals do. They have a large number of similar attributes, share similar feelings, and can even be prepared to do human errands.

There are 5 fundamental types of gorillas and north of 260 types of monkeys. All primates are partitioned into 2 classes; Old World Monkeys and Apes, and New World Monkeys. Old World primates incorporate every one of the gorillas just as people. They are local to Asia and Africa. The New World Monkeys live in the Americas. Old World monkeys likewise never have tails that can be utilized for climbing, while numerous New World monkeys do.

This book will investigate the primates in general, and a large number of the most well known monkeys and offer intriguing realities and striking photos of these excellent, clever animals.

Chapter 1 - Gorilla

Gorillas are the biggest living primates - the group of creatures that incorporates monkeys, chimps and people.

A grown-up gorilla can be more than 6 feet tall and weigh as much as 500 pounds. They can spread their arms 8 feet across and are pretty much as solid as up to 8 people.

Gorillas have 10 fingers, 10 toes, front oriented eyes, ears on the sides of their heads, and 32 grown-up teeth, actually like people do. Be that as it may, in contrast to individuals, their large toes really look like thumbs and help with climbing, and their bodies are covered by thick, dull hair, besides all over, chest, underarms, the centers of their hands and bottoms of their feet.

When a gorilla comes to around 15 years of age, the hair on his back will become silver, and he will be known as a Silverback Gorilla. A Silverback Gorilla is the full grown, experienced male head of a gathering of mountain gorillas in nature.

The Silverback is liable for the security of his gathering of gorillas which can number from 5 to around 30 individuals. These gatherings are called 'troops'.

The Silverback settles on each of the choices in regards to the troop, for example, where it rests, where it searches for food, and where it voyages.

Generally, gorillas are extremely friendly creatures among others in their gathering. They play and communicate, and show a large number of the very feelings that individuals do. Be that as it may, sometimes, a male gorilla from another troop will challenge a Silverback.

The Silverback will beat his chest with his hands, shout and kick up some dust, and charge at the gatecrasher. Silverbacks likewise may sever branches from a tree and shake them at interrupting gorillas.

Gorillas ordinarily stroll on their back feet and knuckles of their hands. They can stand, however don't regularly do as such, except if it is to beat their chests.

Baby gorillas are around 4 ½ lbs when conceived, and are totally vulnerable. They will depend on their mom for their initial 3 years, and families regularly stay together forever.

Gorillas speak with one another by utilizing signals, body stances, looks, vocal sounds, chestslaps, drumming and smells. In spite of the fact that they can't utter the hints of human discourse, gorillas are fit for understanding communicated in dialects and they can figure out how to impart in communication via gestures.

Gorillas are exceptionally insightful, and experience a large number of similar feelings individuals do like joy, dread, misery, satisfaction, love, disdain, pride, liberality and desire. They will giggle when tickled, and cry (in spite of the fact that with sounds, not tears) when harmed.

Gorillas will rest around 12 to 13 hours out of each night, and lay down for incessant rests during the day. At the point when they are not resting, they are normally searching for food.

They principally eat leaves, shoots, natural products, bulbs, bark, plants and bothers. They likewise eat subterranean insects, termites, grubs, worms and creepy crawly hatchlings.

There are three kinds of gorillas; Western Lowland, Eastern Lowland and Mountain gorillas. The names allude to the various spaces of Africa where they reside.

When the Silverback in a troop passes on or is killed, then, at that point, the troop will head out in a different direction and view new gatherings as a piece of. Nonetheless, on the off chance that a more youthful male difficulties the pioneer and wins, he turns into the new pioneer and the troop remains together.

All gorillas are imperiled because of natural surroundings misfortune and hunting by people.

The normal gorilla will satisfy 40 years in the wild, and 50 years in a zoo.

Chapter 2 - Orangutan

Orangutans live in Indonesia and Malaysia on the islands of Sumatra and Borneo. These are the main spots where they reside in nature.

An orangutan is around 4 ½ feet tall and weighs around 200 lbs. They have extremely long arms for their bodies that can extend 8 feet.

Unlike different gorillas, orangutans are arboreal, which implies they invest most of their energy in trees, never boiling to the cold earth.

They rest in a home in the trees made from leaves and branches.

Also dissimilar to different chimps, they like to live alone. A mother might live with a couple of infants, yet guys quite often live alone.

Orangutans are exceptionally shrewd creatures, regularly making themselves umbrellas out of enormous leaves when it rains, and use sticks to get honey from bee colonies.
They eat for the most part organic product, bark leaves, honey and bugs.

An orangutan can satisfy 45 years of age in nature.

In the Malaysian language, Orang implies individual, and utan, (got from hutan) implies from the backwoods.

Like the gorilla, the orangutan's territory is contracting due mostly to deforestation and on the off chance that the pattern isn't switched, the Orangutan living on the island of Sumatra might be the first of the Great Apes to become wiped out.

Chapter 3 - Gibbon

The gibbon is a more modest estimated primate that lives in the wildernesses of Southeast Asia. Gibbons are up to 3 feet tall and just weigh around 15 to 20 lbs.

Gibbons' hands are actually similar to our own; they have four long fingers in addition to a more modest opposable thumb. Their feet have five toes, including an opposable enormous toe. Gibbons can handle and convey things with both their hands and their feet.

Gibbons are omnivores, which means they eat the two plants and meat. They will eat natural product, leaves, blossoms, seeds, tree covering, and delicate plant shoots. They additionally eat bugs, insects, bird eggs, and little birds.

Unlike different chimps, gibbons don't make a bed to rest on around evening time. They rest upstanding in tree limbs.

Gibbons, similar to the orangutan, invest most of their energy in the trees.

When they do boil to the cold earth, they are bipedal, which means they stroll on two legs very much like individuals.

Gibbons move all through the trees by swinging from one branch to another.

This is called brachiating. They can likewise stroll along little branches high undetermined, similar to tightrope walkers; they utilize outstretched arms to assist with keeping their equilibrium.

Each morning when a gibbon stirs, it will declare its quality to different gibbons by making an uproarious, hooting call that goes on for about ½ 60 minutes. This is likewise to tell different gibbons not to attack their domains.

A gibbon will satisfy 40 years of age in nature.

Chapter 4 - Chimpanzee

Chimpanzees are one of our nearest living family members. Truth be told, people and chimpanzees share around 98% of a similar DNA!

In the wild a chimpanzee might live to be 60 years of age and weigh as much as 120 pounds. They live in Africa's thick tropical jungles, open forests, and meadows. They are presently found in 21 diverse African nations.

Chimpanzees convey in numerous ways, generally through sounds and calls. They likewise speak with one another through touch, looks and non-verbal communication.

The chimpanzee is an incredibly keen creature. Truth be told, it utilizes a larger number of devices for additional reasons than some other animal with the exception of individuals.

Chimpanzees are omnivores, eating natural products, nuts, seeds, and leaves, just as numerous sorts of creepy crawlies and periodically medium-sized creatures.

During the day chimpanzees invest the majority of their energy in trees, once in a while swinging from one branch to another. When strolling on the ground, they stroll on their legs and knuckles like the gorilla.

Chimpanzees are exceptionally friendly and live in bunches called 'networks' of up to 100 unique chimps.

Chimpanzees can frequently be seen preparing each other, here and there for a really long time at a time. This is utilized for skin health management, yet additionally for social holding.

Male chimpanzees vie for predominance by slapping their hands,

stepping their feet, hauling branches, and tossing rocks.
A grown-up chimp is multiple times more grounded than a human.

Much like individuals, chimps trade much love, and can likewise be tainted by numerous human infections like measles and influenza.

When they are wiped out, chimpanzees will eat therapeutic plants to mend themselves, and when they are exhausted, they will put together games to engage themselves.

Chapter 5 - Bonobo

The bonobo is basically the same as a chimpanzee. Indeed, it is regularly called the 'bantam chimpanzee' or 'dwarf chimpanzee.'

The main spot on the planet bonobos are found is in the Zaire River locale of The Congo in Africa.

Although they are known as the bantam chimpanzee, they are in reality pretty much a similar size. The normal life expectancy of a bonobo is around 40 years.

The bonobo is such a lot of like the chimpanzee that it was distinctly toward the start of the twentieth century that it was perceived as an alternate animal varieties.

Their eating routine, resting propensities, qualities, activities and propensities are largely practically indistinguishable from the chimpanzee.

Chapter 6 - Baboon

Baboons are enormous and incredible monkeys that invest the vast majority of their energy on the ground since they don't have a 'grasping' tail like some other monkeys.

They are found all through Africa, and can live in pretty much any kind of climate as long as there is a water supply.

Baboons are extremely friendly with each other and live in huge soldiers of dependent upon at least 100.

The primate is an omnivore eating organic products, plants, leaves, bugs, reptiles and rodents. They are so enormous and incredible with amazingly sharp teeth that they will likewise chase after bigger prey like gazelles.

Baboons impart through calls, snorts, barks, shouts, yawns, shoulder shrugs, lip smacking and even tail swaying.

Due to their size and the way that Baboons spend most of their lives on the ground, they are gone after by various hunters like lions, panthers and cheetahs alongside huge pythons, African wild canines and flying predators.
Baboon's bright bottoms have thick cushioning on them to permit them to sit for significant stretches of time.

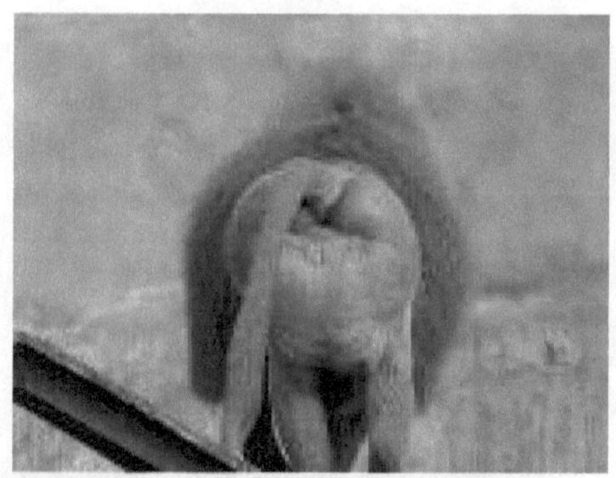

There are five unique types of mandrills and they come in various shadings, both on their bodies and on their appearances.

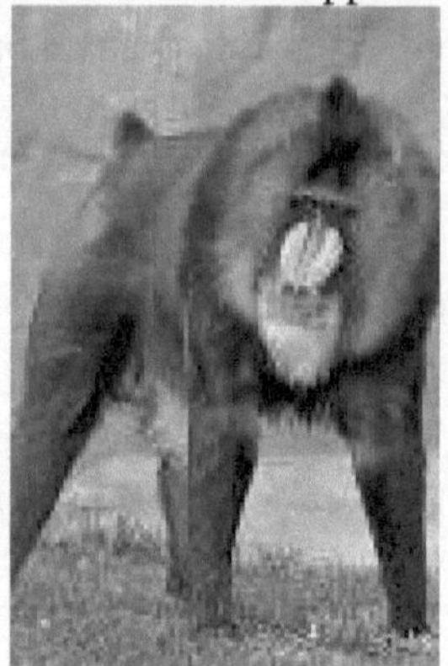

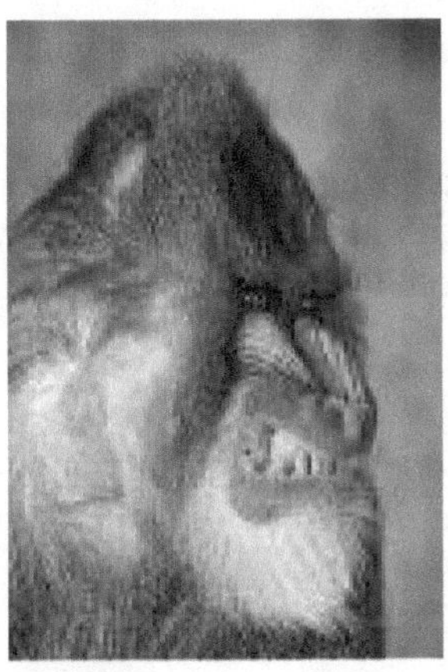

Chapter 7 – Squirrel Monkey

Squirrel monkeys are the most widely recognized monkey living in the tropical jungles of Central and South America.

Squirrel monkey are exceptionally modest, and utter not very many sounds, however may screech whenever terrified or in harm's way.

They by and large stay in the highest points of trees, however come down to search for food.

They like to eat various kinds of blossoms, leaves, buds, nuts, creepy crawlies, reptiles and eggs. They spend right around almost 100% of their time in the trees.

Squirrel monkeys are tiny, standing scarcely a foot tall and weighing under 3 lbs.

Squirrel monkeys spread pee on all fours to stamp their way when they are traveling through the treetops. By following the fragrance, different individuals from the gathering can find one another.

Because of their little size, Squirrel monkeys are gone after by birds and snakes.
Squirrel monkeys are superb jumpers, and move all through the timberland by bouncing from one tree to another, occasionally covering incredibly significant distances.

Squirrel monkeys live for around 15 years in the wild and around 20 years in a zoo.

Chapter 8 – Howler Monkey

If you've at any point heard a howler monkey at the zoo, you most certainly know where they got their name. They are the most intense of all monkeys, and their yells can be heard for up to 3 miles away.

They can be found in the rainforests of Central and South America. They are around 20 lbs and stand 4 feet tall.

Howler monkeys quite often stay in the treetops and feed on leaves, natural products, nuts, and blossoms.

Howler monkeys have a prehensile tail, which implies that their tail is fit for holding. They use it as a fifth appendage to get a handle on branches and move from one tree to another.

The highest point of the monkey's tail is covered with hide, yet the base isn't, which permits the tail to improve hold on the branches.

Howler monkeys live in troops of around 15 to 20 monkeys, and live for around 20 years.

Howler monkeys have whiskers and long, thick hair which might be dark, brown, or red.

The howler monkey will for the most part make its commotions toward the beginning of the day and by the day's end to give others access the troop know where they are at, just as to caution and caution different soldiers not to go into their domain.

The howler monkey is one of the most un-dynamic monkeys in nature. They spend up to 80 percent of their day either dozing or resting in the trees.

Chapter 9 – Spider Monkey

Like a large number of the other monkeys in this book, the bug monkey lives in the rainforests of Central and South America.

They likewise invest most of their energy in the treetops, once in a while boiling to the cold earth.

They are regularly around 4 to 5 feet tall and weigh around 13 lbs.

They have an incredibly solid tail that assists them with swinging from one tree to another. They are unquestionably spry and exceptionally enjoyable to watch. They have been nicknamed the acrobats of the wilderness!

They are extremely friendly monkeys, generally living in gatherings of 20 to 30 monkeys or more.

They can be uproarious creatures and regularly speak with many calls, shrieks and barks.

The primary hunters of arachnid monkeys are panthers, jaguars, ocelots and enormous snakes.

When various soldiers of arachnid monkeys meet, they embrace each other to show companionship and stay away from conflict and hostility.

Spider monkeys don't have a thumb. Their four fingers are bended and assist them with holding tight to the parts of trees.

The insect like development they make while swinging from trees, is the thing that gives them the name bug monkey.

The insect monkey has the most grounded tail in the animals of the world collectively.

Chapter 10 – Rhesus Monkey

Rhesus monkeys are found in Afghanistan, Pakistan, India, Southeast Asia, and China. These insightful creatures can adjust to numerous living spaces, and some can even become acclimated with living in human networks. This is generally normal in India, where Hindus see the creatures as hallowed and for the most part leave them undisturbed.

Rhesus monkeys are great swimmers. Indeed, even babies have great swimming capacities. They additionally have a long tail that assists them with remaining adjusted and utilized for jumping.

The rhesus monkey's eating regimen incorporates roots, organic product, seeds, and bark, and furthermore bugs and little creatures.

They live and are dynamic both in the trees and on the ground.

Because of their nearby likenesses to people, the rhesus monkey is critical to logical examination in regards to wellbeing related investigations.

The rhesus monkey is around 21 inches tall and gauges 17 lbs. They live for around 30 years in zoos, however regularly don't make it recent years old in the wild, for the most part because of hunters like canines, weasels, crocodiles, enormous snakes, flying predators and Bengal tigers. These monkeys love the water so a lot, that they have even been known to swim in shark plagued waters and become a dinner for the sharks.

Rhesus monkeys were the primary creatures sent into in space, even before people. The United States sent a Rhesus monkey named Albert I into space prior to sending human space explorers.

Chapter 11 – Capuchin Monkey

Capuchin monkeys live in the rainforests of Southern Central America.

Capuchin monkeys are strikingly keen and are the most brilliant of the New World monkeys. Capuchin monkeys have regularly been utilized as allies for impaired individuals. These monkeys have hands like human hands, and they can perform helpful activities, such as serving food, opening and shutting the entryway, turning on and off lights, recovering articles required, and brushing hair.

Capuchin monkeys are additionally kept as pets, yet extraordinary consideration is needed for these still wild creatures.

except for an early afternoon rest, they go through their whole day looking for food like natural products, nuts, blossoms, seeds, bugs, insects, bird eggs and little creatures. Around evening time they rest in the trees, wedged between branches.

They are small monkeys, coming to around 20 crawls in tallness, and weighing up to 9 lbs.

Main hunters of capuchin monkeys are boa constrictors, panthers, birds of prey and hawks. They make whistling clamors to caution other monkeys of risk.

They murmur when they welcome one another.

Capuchin monkeys can satisfy 50 years of age in imprisonment.

Because it is so handily prepared thus exceptionally sharp, the capuchin monkey is a well known monkey utilized for motion pictures and TV programs like Night at the Museum, Dr. Doolittle, Zookeeper, We Bought a Zoo, and George of the Jungle.

Chapter 12 – Marmoset

The marmoset is a tiny New World monkey that possibly comes to around 8 crawls in tallness when completely developed.

This monkey lives in the inside of the thick Central and South American woods.

Marmosets are exceptionally crude monkeys and have hooks rather than nails like a portion of the other monkeys.

Marmosets are omnivores, eating predominantly organic products, leaves, tree sap, bugs, bird eggs, bugs and more modest vertebrates.

In the wild, a marmoset will generally live around 10 years, and can satisfy 16 years in a zoo.

One of the most well known kinds of marmosets with kids is the Pygmy Marmoset.

The dwarf marmoset is the littlest monkey on the planet. It is here and there called a 'finger monkey' since it can really fold its whole body over one of your fingers!

They are something like 5 inches long and weigh around 4 ounces. That is about a large portion of the heaviness of the normal juice box!

But don't allow their little size to trick you. Dwarf marmosets are exceptionally deft and can bounce up to 15 feet noticeable all around!

Chapter 13 – Black and Gold Lion Tamarin

The two most normal sorts of tamarin monkeys are the dark lion, and the gold lion tamarins.

They get their names from the manes around their faces, like a male lion's.

They are found distinctly in the rainforests of Brazil, and are one of the most fundamentally imperiled creatures on the planet.

The dark lion tamarin is additionally normally known as 'the brilliant rumped lion tamarin'.

Their size, propensities, diets and qualities are basically the same as their cousins, the marmosets.

Chapter 14 – Colobus Monkey

The colobus monkey is handily perceived by its U-molded white hide running from its shoulders to its upper back, and encompassing its face. There is additionally a red colobus monkey, with spaces of reddish hued hide on their generally dark bodies.

They are the main types of monkey without thumbs. Indeed, colobus signifies 'twisted' or 'docked' in Greek. This is on the grounds that they have a little stub where a thumb would be.

They are found in focal Africa, and can live around 20 years in the wild, and as long as 30 years in imprisonment.

They weigh around 30 lbs, and are up to 24 inches tall, excluding their tail which can be up to 40 extra inches long!

The colobus monkey utilizes branches as trampolines, bouncing all over on them to get force for jumps of up to 50 feet. They jump up and afterward drop descending, falling with outstretched arms and legs to snatch the following branch.
Their hair and tails go about as a parachute during these long jumps.

Bonus – Lemurs and Other Monkeys

The lemur isn't really a monkey or a chimp, yet it is in a similar primate family as they are, similarly as people are.

They are found chiefly on the island of Madagascar on the African coast.

Lemurs are around 18 inches long and weigh up to 8 lbs. Their tails are about one more 2 feet long.

Lemurs live in gatherings of up to 25, and there is no distinct head of their soldiers, in any case, the females do rule over the guys.

Most kids will perceive the lemur as King Julien in the film Madagascar.

Owl Monkey

Gelada

Proboscis Monkey

Saki Monkey

Emperor Tamarin

Snub Nose Monkey

**Drill Monkey

Mandrill